50 Things to Know

50 THINGS TO KNOW ABOUT LEARNING PROGRAMMING

LEARN TO PROGRAM THROUGHOUT YOUR LIFE

Kirollos J.B. Hanna

50 Things to Know About Learning Programming Copyright © 2018 by CZYK Publishing LLC. All Rights Reserved.

All rights reserved. No part of this book may be reproduced in any form or by any electronic or mechanical means including information storage and retrieval systems, without permission in writing from the author. The only exception is by a reviewer, who may quote short excerpts in a review.

The statements in this book are of the authors and may not be the views of CZYK Publishing or 50 Things to Know.

Cover designed by: Ivana Stamenkovic
Cover Image: https://pixabay.com/en/programming-html-css-javascript-1873854/

CZYK Publishing Since 2011.

50 Things to Know
Visit our website at www.50thingstoknow..com

Lock Haven, PA
All rights reserved.
ISBN: 9781791532932

50 THINGS TO KNOW BOOK SERIES REVIEWS FROM READERS

I recently downloaded a couple of books from this series to read over the weekend thinking I would read just one or two. However, I so loved the books that I read all the six books I had downloaded in one go and ended up downloading a few more today. Written by different authors, the books offer practical advice on how you can perform or achieve certain goals in life, which in this case is how to have a better life.

The information is simple to digest and learn from, and is incredibly useful. There are also resources listed at the end of the book that you can use to get more information.

50 Things To Know To Have A Better Life: Self-Improvement Made Easy! by Dannii Cohen

---

This book is very helpful and provides simple tips on how to improve your everyday life. I found it to be useful in improving my overall attitude.

50 Things to Know For Your Mindfulness & Meditation Journey by Nina Edmondso

---

Quick read with 50 short and easy tips for what to think about before starting to homeschool.

50 Things to Know About Getting Started with Homeschool by Amanda Walton

I really enjoyed the voice of the narrator, she speaks in a soothing tone. The book is a really great reminder of things we might have known we could do during stressful times, but forgot over the years.

- HarmonyHawaii

50 Things to Know to Manage Your Stress: Relieve The Pressure and Return The Joy To Your Life

by Diane Whitbeck

There is so much waste in our society today. Everyone should be forced to read this book. I know I am passing it on to my family.

50 Things to Know to Downsize Your Life: How To Downsize, Organize, And Get Back to Basics

by Lisa Rusczyk Ed. D.

Great book to get you motivated and understand why you may be losing motivation. Great for that person who wants to start getting healthy, or just for you when you need motivation while having an established workout routine.

50 Things To Know To Stick With A Workout: Motivational Tips To Start The New You Today

by Sarah Hughes

50 Things to Know About Learning Programming

BOOK DESCRIPTION

Are you interested in learning about programming but don't know where to start?

Do you want to avoid the common pitfalls that self-taught programmers tend to often make?

Do you want to clear your misconceptions about the process of programming?

If you answered yes to any of these questions then this book is for you...

50 Things to know about learning programming by Kirollos Hanna offers a look inside a learning programmer's mind and experiences. Most books on programming tell you to learn the rules of programming without telling what to do with those rules. Although there's nothing wrong with that, knowing what to do after learning the basics can save you a lot of time. Based on knowledge from the world's leading experts doing programming is much better than just learning about it.

In these pages, you'll discover the mistakes most beginners make when they start their journey as well as the many ways of increasing your experience faster. This book will help you know how to achieve your goals related to programming. Whether you want to get hired, build a software service or just improve your skills.

By the time you finish this book, you will have a better understanding of programming as a whole. So, grab YOUR copy today. You'll be glad you did.

TABLE OF CONTENTS

50 Things to Know
Book Series
Reviews from Readers
BOOK DESCRIPTION
TABLE OF CONTENTS
ABOUT THE AUTHOR
INTRODUCTION
1. Programming is a Bunch of Simple Rules That Make Complex Systems
2. Programming Is Mostly Googling
3. Make a Plan to Reach Your Goals
4. It isn't About Syntax, It's About Solving Problems
5. Good Programmers Don't Try to Know Everything
6. You are not Stupider Than Any Other Programmer You've Met
7. Don't Be the One-Programming-Language-Is-Best Kind of Person
8. Your Imposter Syndrome is Slowing You Down
9. Don't Reinvent the Wheel
10. Don't Overcomplicate Things
11. Don't Use Motivation, Use Habits and Consistency
12. Always Plan Before You Code

13. Simplify Complex Problems
14. Don't Repeat Yourself
15. Keep It Simple, Stupid
16. You'll Never Feel Ready
17. If You're Just Starting Out, Pick A Language and Stick to It
18. There is No Quick Way to Learn Programming
19. Copying Is Normal as Long as You Know What You're Doing
20. Making Mistakes Is Normal
21. College Is Not the Only Way into The Industry
22. Landing A Job Is Easier Than You Think
23. Programming Jobs Are About Projects Not Certificates
24. Interview Questions Often Have Little to Do with Your Actual Job
25. Soft Skills Are Important for Programmers
26. Contributing to Open Source Can Improve Your Skills Quickly
27. Don't Forget to Move
28. When Watching Programming Tutorials, Try to Write the Code After the Explanation
29. Be Careful of Tutorial Purgatory
30. Teaching Programming Can Make You A Better Programmer

50 Things to Know

31. There Will Always Be Things You Can Improve in Your Project
32. Change It Up Every Once in a While
33. Read Popular Programming Books
34. Learn to Read Documentation Effectively
35. Make Programs to Solve Real World Problems
36. Make Sufficiently Challenging Projects
37. Make Side Projects to Develop Your Skills
38. Use Coding Sites to Improve Your Problem-Solving Skills
39. Blogs are Great for Learning Concepts Quickly
40. Learn Version Control
41. Test Your Code
42. Debugging Is Most of the Job
43. Refactor Your Code When You Need To
44. Be Confident When Presenting Yourself as A Programmer
45. Find A Mentor and Remember to Ask for Help
46. Participate in Social Activities and Events
47. Celebrate Your Achievements
48. For Web Development, Learn Javascript
49. For Data Science and Machine Learning, Learn Python or R
50. For Mobile App Development, Learn a Native Language for The Platform of Your Choice

Other Helpful Resources

50 Things to Know

ABOUT THE AUTHOR

Kirollos Hanna is a college student with an interest in programming. He has been programming for more than two years in a variety of different programming disciplines. He has made many small projects for his own interests.

He wants to show the world how programming can affect a person's way of thinking and guide aspiring programmers in their learning journeys.

He is currently studying management information systems and programming on the side.
You can find him on twitter @kirojb

INTRODUCTION

"Premature optimization is the root of all evil."
— Donald Ervin Knuth, The Art of Computer Programming

Many people with a desire to learn programming are intimidated by it. They believe that they have to be a genius or a math wiz to be a good programmer. That myth is simply not true and thankfully, it's dying down.

Programming is a more simple form of writing. If you've ever written a sentence before, you're smart enough to program. But the thing that makes it difficult is its simplicity. Our everyday communications with each other rely on a mutual understanding of our language, culture, etc… And the thing about computers is that they don't intuitively understand anything.

A programmer is the one who makes the computer understand what to do.

Computers and technology are allowing us to automate most of the grind that we used to do in the past and there's still more to automate in our lives. But other than automation, we have become much more connected and capable because of technology.

In these pages, we'll delve into many of the misconceptions people have about programming, a programmer's motivations and his/her way of thinking as well as the most common ways a programmer solves problems.

1. PROGRAMMING IS A BUNCH OF SIMPLE RULES THAT MAKE COMPLEX SYSTEMS

When you're starting out, you might think that a programmer uses mathematical magic in order to code. But the truth is that all you have to do is learn a programming language and write lines that tell the computer what to do. It's very similar to normal writing, if not even simpler in nature-- You won't have to worry about how to pronounce the word "read", for example.

Programming languages are built on a bunch of simple concepts (i.e. logic, iteration, abstraction…) that allow programmers to make all the programs that we use on a day to day basis. From the software in your washing machine to the screen you're reading this on—everything is built with the same programming languages.

2. PROGRAMMING IS MOSTLY GOOGLING

As a Programmer, you'll be surprised at how much you're going to use Google. Don't think of it as cheating, because Google has become a de facto handbook for every programmer out there. Everybody uses it—Every day.

With more experience in the field, you'll be able to google better and that's all that separates a junior programmer from a senior programmer. Finding the right questions to ask the search engine is often the hardest part of the job.

3. MAKE A PLAN TO REACH YOUR GOALS

Getting into the programming world can be overwhelming. There is a ridiculous number of fields for a person to choose from and every field requires you to develop skills that may have little to do with programming but they are all connected by programming.

Data science requires you to know statistics and mathematics, UI design requires some knowledge in the psychology of people as well as learning about color theory, fonts, and so on, not to mention the amount of creativity involved. Then there is also game development that is an accumulation of multiple disciplines ranging from animation, art, artificial intelligence, design etc..

If you just tell yourself that you want to learn programming without knowing what you specifically want to learn, you'll end up wasting a lot of time with very little to show.

Make yourself a three-month plan where you detail to yourself what you plan on learning and implementing within this time period. Look for specific courses you want to take in the field of your choice and research some simple projects that you can implement with the knowledge you'll get from those courses.

There is an increasing number of free online curriculums for specific fields in programming that you can join. For example, if you want to dip your toes into web development, look no further than

Freecodecamp.com which is a free website that teaches people web development through practical lessons and projects. Many people have been able to get hired by using the website as a primary learning resource.

Making things is a very important part of your learning journey and with a learning plan, you'll be able to shorten the time gap between just learning to actively making things using your skills.

4. IT ISN'T ABOUT SYNTAX, IT'S ABOUT SOLVING PROBLEMS

Syntax is only a small part of programming. Programming is about solving problems not memorizing syntax. If you want to make an app to keep track of your time. You can learn the syntax of a language like Java or Swift while you make the app for the mobile platform of your choice.

You shouldn't spend hours and hours learning about concepts related to those programming languages and mobile platforms when you don't even know if you're going to use those concepts in your project. The only way to know what you'll need is by actively asking about what it takes to build what you need and not by learning all the concepts of a technology in the hopes of finding the right one for what you want to do.

There is a subtle difference there that you'll get to identify better the more time you spend programming.

Making projects that you don't know how to technically do yet is the best form of active learning for a programmer, and it can also beef up your portfolio. But don't mistake not knowing syntax for doing a project without a plan of action. You'll always need to plan for what you're building.

5. GOOD PROGRAMMERS DON'T TRY TO KNOW EVERYTHING

You can never know everything about programming. That is a fool's errand. More libraries and tools are being uploaded for our use at this very moment than you can ever learn.

Instead, you should know enough to do what you set out to make. Programmers around the world build tools for others to use and make our lives easier. Those tools are what are called black boxes, they provide an expected output for every input without the overhead of knowing how they're done or making them yourself from scratch.

You simply have to learn to provide a specified input and know what the output is going to be for the functions in those black boxes. You don't have to look under the hood of those technologies and learn them inside out to use them, the same way no car driver has to necessarily know how an engine works.

6. YOU ARE NOT STUPIDER THAN ANY OTHER PROGRAMMER YOU'VE MET

While you're on your journey, you'll meet a lot of programmers who seem to know it all. They'll somehow know everything that you don't know.

Don't be discouraged by this, these people have spent countless hours to get to that level of knowledge. And even then, they still won't know nearly as much as you might think.

We're only trying to improve our skills to achieve what we want. Some of us are just further down the road than the rest.

The misconception that programmers are geniuses is being proven to be false every day. There are many stories of people with little education who've turned their lives around by learning programming through the internet. It just takes hard work and time to reach a high level of competence

7. DON'T BE THE ONE-PROGRAMMING-LANGUAGE-IS-BEST KIND OF PERSON

There are people out there who swear by their programming language. Saying things like "Python is the best programming language for everything" or "C++ is the only real programming language".

Don't be like these people.

Every programming language in existence was made for a reason. A language like Python is a popular choice for data scientist, Javascript is a must learn for web development, Java is the go to language for enterprise level applications and C++ is very popular for systems and game development.

At the end of the day, languages are tools. They are a means to an end to help us achieve our programming goals.

No one programming language is best. They all have their own uses and functionalities.

8. YOUR IMPOSTER SYNDROME IS SLOWING YOU DOWN

The feeling of not being good enough to do this—that is imposter syndrome. You'll feel it when you've just started learning about this stuff, it will manifest itself as a voice in your head that tells you how you'll never be able to learn all these things, that only someone more intelligent than you can be competent enough to learn to program.

You'll feel it when you finally manage to finish your portfolio project and you'll hear it point out all the faults in it and how it isn't good enough of a project for you to be proud of.

You'll feel it when you get a job in the industry and see all your coworkers who feel like they know everything that you don't.

The truth, however, is that everybody has this feeling. Your coworkers in your new job have this feeling on a daily basis and they probably know what you're going through as well. That project you made and finished is an accomplishment because 90% percent of people who start their own projects never finish them. And everybody who has ever starting learning about programming felt they were incompetent.

You have to deal with the fact that your imposter syndrome may never go away, instead you can use it to fuel your desire to improve in your craft. We all have to deal with it and everybody has their own way of doing so but you must not let it stop you.

9. DON'T REINVENT THE WHEEL

When you're trying to implement a new feature, try to see if someone has already done it before. There are many free frameworks and features that programmers can use and modify to fit their own needs without them having to rewrite a lot of the code.

Use these tools when you need them, they will save you a tremendous amount of time and energy.

You might feel a bit of indignation at first about using other people's work in your own projects but remember that computer technology didn't get this far in 60 years by having every programmer do the same things twice to make sure it's their own work.

People use other's work because it's less time consuming and allows for more productivity and improvement in the long run. Use the tools available to you and improve upon them, making things that have already been done before is no fun.

10. DON'T OVERCOMPLICATE THINGS

Many times when you're starting out, you'll want to do things the "right" way. You'll look everywhere on the internet for the "right" way of doing such and such but you'll only find things that may not be efficient enough for your standards.

If this is the first piece of code you're writing for your program and you're not trying to increase the efficiency of your programs then use the best solution that you can find and don't wait for the perfect answer to your particular problem.

If you have a solution to a problem that you're facing, use it. Don't think about its efficiency for now. That should happen when you're optimizing your program.

The problem of overcomplicating your coding process can happen because of the incredible amount of options we have as programmers of doing the exact same thing in different ways.

While there are indeed more efficient methods than others, you don't have to find them right away or else you won't be satisfied. Such a way of thinking is very paralyzing for programmers.

11. DON'T USE MOTIVATION, USE HABITS AND CONSISTENCY

When you're just starting out, you might think that it might be a good idea to study eight hours a day, every day, until you become this superstar programmer. Do yourself a favor and don't try to do that.

It's important to improve your skills everyday there is absolutely nothing wrong about that. What is wrong, is thinking you'll last. Our minds don't adapt to such rapid changes in our daily habits easily. Just because you're motivated to do so today, doesn't mean you will be tomorrow.

Remember to start your programming habit slowly and take things one step at a time. Being consistent is much more important than pumping out eight hours of programming in one day and then stopping for a few weeks because your brain is exhausted.

12. ALWAYS PLAN BEFORE YOU CODE

Planning your code will give you a road map that you can work towards. It will allow you to measure progress and therefore, stay motivated more easily.

You don't have to follow your plan to the letter, but having it there will be better than coding blindly. Use your plan as a reference that you check every time you implement a function or a feature. Always ask yourself if you're doing too little or too much in regards to the specific functions and features that you want in your project.

The best part about having a plan is that you'll be able to convince yourself to move on to the next part of your project if you feel like what you did is more than what you planned for. As a programmer, you might get too fixated on a specific part of your project with the intention of making it perfect, which is a big waste of time in the long run since nothing can be perfect and you'll always want to do more.

On the other hand, you might have not given a feature its due diligence and decide to move on too soon but your plan will serve as a deterrent in this scenario.

13. SIMPLIFY COMPLEX PROBLEMS

If you want to avoid a meltdown, don't tackle big complex problems as a whole. Break them down bit by bit until you see something solvable and then start working on a solution.

How you break them down mostly comes with knowledge and experience. You have to know the problems that your computer can solve so that you know what to simplify your problems into.

On the other hand, if you've been trying to solve this one bug in your code for what feels like ages now and you feel like you're about to explode from frustration, acknowledge that it's okay and realize that it happens to everyone.

Take a break and go do something else. Come back to the problem later with a fresh mind. It will do wonders for your mental health and your problem-solving skills.

14. DON'T REPEAT YOURSELF

There is a popular concept that programmers tend to live by called DRY code (DRY stands for don't repeat yourself). Every block of code you write should do its own thing. Never write code that does the same thing twice.

This is because it gets very hard to maintain such code in large projects. Since if you want to change the function of this code, you'll have to find all instances of this code and change it one by one.

While this may sound simple, it could be tricky to master. Because we don't always realize how much repetition we make in our own code. Don't be discouraged, however, since with practice comes mastery.

15. KEEP IT SIMPLE, STUPID

Keep it simple, stupid or KISS is similar to DRY code. When you're writing a block of code and you have the option to choose between writing a complicated one liner and multiple lines of simple code that do the same thing, always choose to write the latter.

This concept is particularly important when you're working on a job or in a team. The reasoning behind it is that you won't be the only one working with this block code, just because you know what a complicated one liner means doesn't mean that every person on your team does too.

If a team member who's working on this code base with you sees this complicated one liner they might end up wasting precious time looking up what the line does or maybe if you work in a big company, someone from another department will ask HR to tell them who wrote this line of code because they don't understand it and will eventually go and ask you what it does and so on…

This whole ordeal has actually happened with a huge number of people in companies and it can easily be avoided if you choose to KISS.

16. YOU'LL NEVER FEEL READY

There is this misconception in programming; people think that learning programming is a linear process that you go through the same way you go through learning a subject in school.

However, the truth of that matter is that you'll never feel competent in your skills. You can learn new things every day and still not feel good enough. The secret to being a good programmer is to learn just enough to be able to do something with your knowledge.

Whether you're planning a big project or just want to get a job in the industry, you'll never feel ready. So, start working on that project and learn along the way or apply to that job you've found and get rejected. Do the bare minimum to move forward towards your goal and then once you've reached a good spot improve upon your foundation.

You'll be surprised at how much faster you'll reach your goals if you start without feeling like you're ready.

17. IF YOU'RE JUST STARTING OUT, PICK A LANGUAGE AND STICK TO IT

Sticking to a language at first, will help you improve your skills faster. You can then transfer those skills and experiences into other languages.

However, if you're just learning languages without using them to build or solve something. Then you'll eventually end up learning and relearning the same concepts using different syntax and your knowledge won't stick.

Remember that at the end of the day languages are just tools that help you build things. Some languages are better than others for specific fields. However, when you're starting out, any language can be versatile enough for you to explore your interests and still be able to make simple versions of them without relearning the same concepts in another language.

Unless you're learning something like C++ for web development or PHP for system development which is a very nonsensical choice to begin with. Know what a language is best used for and dive into it, make the best of what you learn and then when you feel you're no longer a beginner, which is when you've made something big enough to can get you job, you can then learn other languages and pursue more interests.

18. THERE IS NO QUICK WAY TO LEARN PROGRAMMING

Even if you're just starting out, you might've already come across books or blogs that promise you that you can learn a programming language in 21 days or a very technical programming subject in a week.

While you may actually be able to be immediately productive in a programming language that you've never used before, that can only happen if you've already studied some previous programming language that is similar to the new language you're using and you have extensive knowledge and experience in programming that spans years of experience.

Nobody, no matter who in this world, can learn a previously foreign looking programming language in a short span of time. In fact, nobody is ever done learning a programming language, because they evolve through time and with their improvement, it comes down to the users of this language to start learning more. The learning is never finished, you just have to know enough to be productive.

As with everything else in life, it takes time to learn something new. Instead of obsessing over how fast you can learn a subject in programming, set a few hours, or even minutes, each day and dedicate this time to learning this subject. Stagnation in your learning may happen, but that is natural.

Should you feel frustrated when you've stagnated, just remember to not give up and finish your studies for the day according to the time you've dedicated. You'll feel the progress through time but if you give up, you'll never improve.

19. COPYING IS NORMAL AS LONG AS YOU KNOW WHAT YOU'RE DOING

One of the most popular sites in the field of programming is called Stackoverflow. It is a website where people can post their technical questions and receive answers from a massive community of other programmers.

The website is also famous among its users because people tend to copy the code in the answers and use it for themselves. People who are new to this might think that such a thing is unethical or somehow wrong.

But it is far from wrong, in fact it's sometimes encouraged. Copying code saves you time and energy and you're still using it for your own projects that have nothing to do with other people's projects. The whole thing is similar to a writer copying a few words into his story. Nobody would think such a thing is wrong since it fits into a different context.

However, the problems tend to arise when you copy mindlessly. When you don't know what the code you're copying into your project means. Doing so will hinder your learning experience and may lead to a lot of bugs down the line.

When you copy code, read it line by line first and understand what it does, so that when something breaks, you'll be able to find the errors in your code and examine the code you copy critically.

20. MAKING MISTAKES IS NORMAL

In programming, mistakes are the name of the game. If you're not making mistakes, you're not learning. This power comes from the fact that programmers can always undo what they did wrong.

They get to learn from their mistakes in practical ways that few professions can do. A doctor can never make a mistake on a patient or else they will risk their death. But programmers can always disfigure their programs and just Ctrl+Z out of the damage once they learn the cause of the problem.

That way, they learn faster through their mistakes and improve quicker.

Some people even argue that not making enough mistakes when you code, hinders your learning experience and makes you stagnate.

21. COLLEGE IS NOT THE ONLY WAY INTO THE INDUSTRY

Programming is a craft like any other. Don't let anyone tell you that it's only for exceptional people who happen to have degrees in computer science or mathematics. There are now more free resources to learn to code than ever and the industry itself is changing faster than college programs can keep up with.

Even if you do have a degree in computer science, you'll likely learn more on the job than in college. Unless you joined a very exceptional college that teaches in-demand programming skills.

Those with no college education or CS degrees don't need to be worried. Simply learn enough to get your foot into the industry and you'll learn more on the job to rival any college graduate.

22. LANDING A JOB IS EASIER THAN YOU THINK

Programmers are needed in every field these days because of the widespread use of computer technology in almost all industries. You can easily jump start your career in programming by doing simple projects in your target niche and applying for jobs or internships.

Making projects shows companies and people that you are interested in the field, can pick up on emerging technologies and can be mentored into a competent programmer. They would, therefore, be more motivated to hire you for your skills in this way.

The hardest part is getting yourself out there but even that is getting easier because of websites like Linkedin, AngelList and other hiring websites that are on the web today.

23. PROGRAMMING JOBS ARE ABOUT PROJECTS NOT CERTIFICATES

Many recruiters have stopped caring about certificates nowadays. Instead, they want proof of your abilities. The market for programmers keeps changing at a break-neck pace and computer science university programs are barely keeping up.

If you can show people what you can do, that's all the qualifications you'll ever need.

To show people that you qualify, you'll need a portfolio with your pet projects. Make at least three small projects and put them in your portfolio, then start applying for jobs. Even if people still reject you, you're likely going to get feedback on why you weren't accepted.

Use that feedback to improve your portfolio and yourself to be better the next time you apply for a job.

24. INTERVIEW QUESTIONS OFTEN HAVE LITTLE TO DO WITH YOUR ACTUAL JOB

Don't be intimidated by the complexity of interview questions. They are meant to show that your problem-solving skills are sufficient for a programming job. However, many people have argued that these interview questions have little to do with the problems you will be facing in your day to day job.

If you're getting a web development job, for example. You'll be working with pre-made functions and APIs most of the time. It is very rare for a web developer to solve a problem that has never been solved before, while the interviews may include hard problems that involve writing dynamic programming code or something of the like and you'll never have to face these problems in your job because someone has already written a library for such a thing.

Whether interview questions matter or not in your job, don't let your results in them dictate your competency as a developer. They are just a gateway for a job-- simple as that.

25. SOFT SKILLS ARE IMPORTANT FOR PROGRAMMERS

Your technical skills won't do you any good if you can't show them to people. Learn how to market yourself, how to effectively sell your services and how to communicate well with potential clients.

These skills have little to do with programming itself but they are the way for us to get our foot into the tech industry.

Putting yourself out there into the world can be one of the hardest challenges you'll have to face, not just in programming but in any field you want to work in. We don't like to be judged by other people and that's normal.

By learning to communicate with the people around us, we'll feel more confident in ourselves and improve the lives of those we connect with.

When you finally set foot into the industry, one of the first things you have to do to maintain a stress-free work environment is to have clear communication with your supervisors. You should know exactly what they want from you, how much freedom you have in making a project and so on.

Your supervisor or manager should also know your boundaries. Some managers may ask too much of you, thinking that it's fine with you. Don't be shy when you tell them politely that they are intruding into your personal space or time.

You'll be thankful for it, in terms of stress-free days. Your productivity will be optimal and your work will be good. Don't put too much stress on yourself and remember that this is a job that you may spend a significant amount of your life in—You don't want to spend that much time under constant stress and anxiety.

26. CONTRIBUTING TO OPEN SOURCE CAN IMPROVE YOUR SKILLS QUICKLY

You might not have heard of the term open source before. If you haven't, you should.

Open source software is software that allows you to see under its hood. In other words, its "source code" is available for you to see and in some cases modify and redistribute.

There are multiple websites that host open source software. The most popular one as of this writing is called Github. Newbie and experienced programmers alike host their projects there. Even huge corporations like Google and Microsoft host many of their open source projects on Github.

These projects allow programmers to contribute to them by writing code that improves said projects. Doing so can not only improve your skills but also gets you noticed by recruiters and big corporations.

27. DON'T FORGET TO MOVE

Programmers mostly work on a desk, so don't forget to exercise and move your body. Focus specifically on your posture and back and do exercises that target those areas.

The strain on our backs while sitting should not be taken lightly. It could affect your body in irreversible ways as you grow old. Buy a standing desk if you can afford it, it will save your back and posture.

Even when you're on a good streak while programming, don't forget to step away from your computer for a little while. Stretch your muscles and stay active. There apps and exercise programs that allow you to energize your body in less than ten minutes. These programs not only reverse the effect of sitting on your body but they also improve your strength and flexibility.

28. WHEN WATCHING PROGRAMMING TUTORIALS, TRY TO WRITE THE CODE AFTER THE EXPLANATION

Programming tutorials tend to show a person writing code while explaining. And while this may not be a bad thing in general, if you just copy the code after them then some of the concepts could easily get over your head.

Take it slow and watch the tutorial multiple times if you have to. And if you can, write the code after the explanation is over. This will show that you really know when to apply this concept in your code.

Don't forget to do all the exercises in the courses you take to consolidate your knowledge.

29. BE CAREFUL OF TUTORIAL PURGATORY

Tutorial purgatory is another popular term among programmers. It refers to a cycle of actions that tend to repeat themselves over and over and end up wasting you a tremendous amount of time.

Many times when a programmer faces a challenge when they are doing a project, instead of googling the answer (as any programmer should normally do) they assume that they don't know enough about the problem and take a 10 hour long course on this problem until they know how to solve it.

Unfortunately, after they solve this problem, another one crops up. So, they buy another course or look up another Youtube tutorial and waste tons of hours to solve a simple problem. By the time they solve this problem, they might even ditch the whole project and think that they are in over their heads.

It's a cycle that keeps repeating itself endlessly. It stems from the fact that some people tend to get obsessive about understanding every aspect of a problem or programming language.

The only we can combat this urge to know everything, is to acknowledge that we can't. Not a single programmer on this planet will ever know everything about programming and those who make great projects don't do them alone. They ask for help from people who can fill the gaps of their understanding with their own.

Know enough about a problem to solve it. It's that simple, but some of us need to get reminded of this sometimes… and that's ok.

30. TEACHING PROGRAMMING CAN MAKE YOU A BETTER PROGRAMMER

The best way to learn a concept is to explain it to someone else. When you teach people programming, you'll notice the gaps in your knowledge better. That way you'll be able to fill those gaps and have a better foundation for learning more in the future.

You'll also become a better speaker or writer depending on your method of teaching which will improve your soft skills and make it easier for you to communicate your thoughts in a clear and concise way when you need to.

There are people who teach programming for a living and most of them started teaching to improve their own knowledge.

31. THERE WILL ALWAYS BE THINGS YOU CAN IMPROVE IN YOUR PROJECT

Your projects are only done when you stop improving them and there will always be things that you can improve, features you want to add, bugs you need to fix….

There comes a point where you have to tell yourself that this is enough for now. Make sure to push your project into production and then add more to it if you want.

To identify that point, it's important to plan your project beforehand and review your plan often.

Whenever you feel like you're going overboard, push your current improvements into production first.

32. CHANGE IT UP EVERY ONCE IN A WHILE

Just because you're a web developer doesn't mean you shouldn't try doing data analysis. Do what you feel like doing. Programming is this other domain of existence that has all these options and variety to it. You can make artificial brains and call it A.I. or design website and mobile applications for the digital world and enrich the lives of those who use the internet and their phones on a daily basis.

You'll never be the master of it all. But then again, you don't have to. You simply have to enjoy what you're doing.

Remember that specializing will get you a job, but the fun in programming will be when you study a bit of everything-- or rather a bit of what interests you. This will keep your passion for the craft going beyond just your job.

33. READ POPULAR PROGRAMMING BOOKS

They are popular for a reason. Every programmer can benefit from the knowledge in those books. Books like "clean code" allow you to write better more understandable code and "Introduction to algorithms" or CLRS is one of the most popular books on algorithms and data structures.

Find the most popular books in the programming topic of your choice and start reading them. Do the exercises and implement the knowledge you gain in practical programs.

34. LEARN TO READ DOCUMENTATION EFFECTIVELY

Documentation is like the dictionary or handbook for technology. If you're looking for a function that does a specific thing in that specific language, you'll be looking for it in that language's documentation.

Nobody knows everything about a technology stack from the get go (duh). Which is why documentation was made.

You'll find documentation for the programming language you're learning, the new framework you're using or the huge number of APIs that you'll end up integrating into your projects.

There is no point in knowing these things like the back of your hand. It would certainly be nice to know them. But most likely, by the time that happens, the technology will already be outdated and you'll be using a different one.

And so, the cycle continues.

35. MAKE PROGRAMS TO SOLVE REAL WORLD PROBLEMS

Any program you make throughout your journey should solve something. If you're just making a program for the sake of making one, you may likely never finish because of your lack of focus and motivation.

On the other hand, if you have an audience in mind, you'll be much more likely to finish your program or project because you know who is going to use it and for what.

Look for the problems you and those around are having in their daily lives and see how your knowledge can make an impact. It might take a bit of creativity on your part but you can always ask for help from those with more experience.

36. MAKE SUFFICIENTLY CHALLENGING PROJECTS

It takes experience to choose the right project for your skill level. If you choose a project that's too hard you'll likely become frustrated and never finish it. If it's too easy, you won't learn anything or improve.

Choose to make projects that you know how to make 70% of-- the rest you should learn while making it. This will help you build upon your knowledge base and improve your experience in using your current skills simultaneously.

However, the challenges in your project don't always have to be technical. Remember that at the end of the day, you're making programs and projects that should be used by others. Understanding the needs of your audience when making a project can also be a very tough challenge that you'll never stop knowing more about.

37. MAKE SIDE PROJECTS TO DEVELOP YOUR SKILLS

Try to always work on a side project in your free time. It'll improve your portfolio and skills. If it's done right, you could even share it with people and gather a user-base. Who knows, your project might become popular and you could monetize it.

Side projects also give you the freedom to choose what you make and how you make it. This is great for your motivation to finish it. But be careful of analysis paralysis, you don't want to continuously plan your project and never actually get to making it.

A huge number of aspiring developers and programmers tend to get stuck on gathering and protecting their ideas without realizing that the implementation of those ideas is the most important factor when it comes to the success of your project.

38. USE CODING SITES TO IMPROVE YOUR PROBLEM-SOLVING SKILLS

There are plenty of coding sites that help you in studying algorithms and data structures on the internet. These sites are known to help programmers in preparing for coding interviews.

They also boost your confidence in your problem-solving skills and familiarize you with multiple programming languages and paradigms.

You can even join coding competitions and win prizes, if you study hard enough. Many people who won competitions have received job offers from famous companies all around the globe.

However, don't think that all the skills you gain from coding sites are directly transferable to any programming job. Just because you know how to implement a certain algorithm and data structure doesn't mean you can clean data or design a website or use an API.

A lot of beginners have the false belief that being an extraordinary competitive programmer makes you a better programmer in their target field which is quickly dispelled when they enter the field that they want. Being a competitive programmer and solving problems on coding sites makes you a generally better programmer but not all your skills transfer into a job.

39. BLOGS ARE GREAT FOR LEARNING CONCEPTS QUICKLY

If you're stuck in a project or just want to brush up on your programming concepts, you should read popular blog posts on those subjects. There are many famous blogs out there for every field in programming.

Pick a few of your favorite blogs and check them for posts that might help you with any specific problem you're having. Be sure that the blogs you're checking are related to the field you're in.

40. LEARN VERSION CONTROL

Version control may sound esoteric to those who hear of it for the first time. But with every project you make, you'll see a pattern of wanting to return a feature that you thought was useless or seeing an earlier state of your program that didn't include this bug that only started happening at a later point in time-- That is when version control comes to the rescue.

By adding versions into your program where every version has particular code for the program. You'll be able to use the command line, a tool that makes you look like a wizard on the computer, to return to earlier versions of your program and seeing the changes you made.

The most famous tool for version control is called git. It is used by beginners and advanced programmers alike. If you don't want to pull your hair out every time you introduce breaking changes to your programs and projects, you'll want to learn about git.

The website Github uses git as a foundation for its services. By using the website, you can collaborate with other developers on the other face of the earth and build a project together. Not only that but you'll also be able to work on separate features on you project simultaneously because of Github's and git's features and services.

You'll also be able to share your programs, projects and tools for other developers to use in their own projects. Open-source programs are very famous for making the lives of developers easier.

41. TEST YOUR CODE

You might not understand how to test or what to test in your code yet but the importance of tests in code must not be overlooked. Testing gives you confidence that your program is behaving correctly.

When you test a block of code, you gain confidence in its output. You know what it will output, therefore you won't need to look back at it if you're program crashes from an unexpected bug. You'll end up saving your time, instead of looking back and re-reading this block of code to make sure it's correct, you'll know that the fault is somewhere else.

There is a paradigm of testing called "Test-Driven-Development" that has taken off in the last few years. Test-Driven-Development or TDD for short is used to give you an outline for your program, similar to how a writer outlines her book before writing it and fleshing out the details.

In TDD, you write tests before actually writing any code which allows you to clear your head when you write the code and gives you a roadmap of how the code that you're writing should behave.

TDD also removes second guessing, which tends to happen as you decide to change something before you've even built it and that in itself is a recipe for disaster since you don't finish what you started and you keep changing the requirements you set for yourself when deciding to build your project.

42. DEBUGGING IS MOST OF THE JOB

Debugging is the act of fixing the faults in your code. When your code is not behaving correctly, you have to "debug" it (i.e. fix it).

You'll find yourself fixing your code more than writing new code. And when that happens know that that's the whole thing about programming. You write something and then you fix it until it works correctly. Everybody does it!

The joy comes when you finally debug what was broken and watch the magic happen.

One famous method of debugging is called the rubber duck technique.

This is a funny way of debugging that a surprisingly large number of programmers use. The way it works is that you have to buy a rubber duck and put it in front of your monitor. Now, whenever you're struggling with bugs, pick the duck up and explain to it what every line of your code does. You'll quickly find the errors in your reasoning and be able to pinpoint the bug. It doesn't even have to be a duck, you can use any sort of toy or even a coffee cup of like. The essential part is explaining your code out loud.

Strangely, this has helped a lot of people in debugging throughout the years.

43. REFACTOR YOUR CODE WHEN YOU NEED TO

Sometimes, you're going to mess up so bad that you're going to have to delete a bunch of old code and replace it with better code. This process is called refactoring and it's done all the time in programming. It doesn't only happen when you mess up, sometimes you have to refactor old code from libraries that you're using in your project or even entire frameworks.

When a piece of code doesn't work and you can't fix it, simply delete it and make a better version of this code.

Refactoring is mainly used to improve a program's performance without changing its output. Since coding languages are always being improved upon, sometimes some functions can become obsolete and newer ones get introduced. At times like these, refactoring is done to maintain functionality and improve performance.

44. BE CONFIDENT WHEN PRESENTING YOURSELF AS A PROGRAMMER

Many people wonder when is the time to call themselves programmers. The answer is when you've made some very simple programs.

Sure, a "hello world" may be too simple but something like a timer or a to-do list, which are the generic programs that most people tend to make, are sufficient for you to know more about programming, than 90% of the human population. Such programs prove your competence in several key areas of programming, they are the minimum amount of knowledge that you must know to be a programmer.

Telling people that you're a programming will not only increase your confidence in yourself but it could also land you jobs from people who need someone with your skillset.

When people hear the confidence in your voice, they'll tend to believe you and if they also have a problem that you can solve with your programming skills, they'll most likely come to you to ask for help.

Confidence in your identity as a programmer also gives you a boost in self-esteem. If you're ever doubtful about yourself when programing, remember that nobody knows everything about programming and there are many senior developers and programmers who google basic syntax in their work every day.

Just because they don't know the syntax or the information required for their profession doesn't make them doubt themselves because they've been through this process before and they know that they will eventually figure it out as long as they keep working and improving.

45. FIND A MENTOR AND REMEMBER TO ASK FOR HELP

A lot of your struggle as a beginner could be avoided if you have a good mentor. When you're programming alone as a beginner, you'll tend to spend countless hours on simple problems that you don't know the answer to.

While such a thing is completely normal and happens to every programmer, you can easily bypass this stage if you have a good mentor who's willing to teach you when you get stuck.

Mentors can be found in companies where you work, or online. Look for experts in the field you're interested in on websites like twitter or join a Facebook group that helps beginners.

You'll be surprised at the amount of help you'll get.

It's okay to ask for help from your friends or mentors when you don't know something, particularly if you have a job. Senior developers in your job have a responsibility of helping juniors.

The industry was built around the mentality of helping each other out. While many of the questions you have can be answered using google these days, you'll still need someone to help you find the right questions to ask.

46. PARTICIPATE IN SOCIAL ACTIVITIES AND EVENTS

People don't like doing things alone, which is why coding groups exist. Most likely, you aren't the only person in your area learning programming. Look for coding groups near you on the internet and go.

Coding groups are a great motivator to keep programming when you're in a slump and don't feel like doing it, not to mention the team building qualities that you acquire from those around you.

If there aren't any near you. Think about making one for yourself and advertising in your area. Start with a small group of friends and go from there. You'd be surprised at the amount of people that might be happy to come. This can also develop your leadership skills.

Hackathons are also great places to meet people who are as enthusiastic about programming as you. People can collaborate together to make things that challenge them and improve their team working skills as well as their technical skills.

Unfortunately, hackathons are few and far between. If you are unable to go to a hackathon that is quite okay, there are other options that will allow you to collaborate with others through the internet.

47. CELEBRATE YOUR ACHIEVEMENTS

Every bug you solve and feature you add, you improve your skills a bit. Celebrate those moments and know that you're getting better at doing your craft.

Programming can be pretty boring without the thrill you get when you manage to find a solution to a problem you have. That dopamine rush you get can motivate you to continue working and improving in ways that usual motivation can't.

Find your own way of celebration. It could be a simple "Yay!" or maybe a little dance, make it physical so your body knows when it's getting rewarded. Another way is celebrating with others, maybe you have a close friend or someone who's going on this journey with you that you share your achievements with.

If you plan on spending your time on programming, no matter the field, you have to make sure it is enjoyable because nobody wants to do unenjoyable things. Celebration is one way of making the process visibly enjoyable.

50 Things to Know

48. FOR WEB DEVELOPMENT, LEARN JAVASCRIPT

There are some programming languages that are known for their use in a particular field. The most popular in this day and age is perhaps Javascript for web development. If you happen to be interested in web development then you've definitely heard about the language already as well as HTML and CSS.

While HTML and CSS are not technically programming languages, they are the languages web developers and designers use to show things on a web page. Javascript, on the other hand, is what provides the web page with interactivity and allows developers to make all the awesome things we use.

You might have also read about all the hate that Javascript gets on the web and that its "bad" or "incomplete". The people who share the hate for Javascript most likely are just used to other languages like C++ or Java that have other uses in programming and the fact that Javascript has very little syntax for the programming paradigms of the other languages makes it feel "messy".

But remember that Javascript was built for the web and will always be there for the foreseeable future. Even with the emergence of technologies like Web Assembly that will allow developers to use other programming languages on the front-end of the web, Javascript will not go "extinct". It is not an exaggeration to say that the whole web's interactivity so far has been built with Javascript. Another thing to keep in mind is that Javascript is evolving, with new iterations of the language being released and supported by browsers every other year or so.

In the end, Javascript is a must learn for web developers and designers of all kinds.

49. FOR DATA SCIENCE AND MACHINE LEARNING, LEARN PYTHON OR R

Python has gained fame in data science because of the extensive amount of libraries supporting data science in the language. On the other hand, R was built for data science and has many concepts that help when doing data science projects built into the language itself.

While the language choice may seem significant, it doesn't actually matter in the end. Languages are just tools and Python and R are very similar in their use for data science—one is not better than the other. It is simply a question of taste for people.

50. FOR MOBILE APP DEVELOPMENT, LEARN A NATIVE LANGUAGE FOR THE PLATFORM OF YOUR CHOICE

Mobile apps have gained a lot of popularity in the past decade. Mobile phones themselves have integrated themselves into our lives in unimaginable ways. The most popular platforms for mobile phones these days are IOS for iphones and apple products and the android operating system for mobile phones from companies like Samsung, Lenovo, Sony, etc…

Another platform for mobile is windows for windows phones, but these have very low popularity and traction as of the time of this writing.

For Android, the native apps are built using Java or, more recently, Kotlin. If you don't know what the difference is between native apps and other apps, it's simply a difference of available tools and performance speed. In this case, other apps include things like games that might be built with a game engine like Unity or apps built with technologies like React Native.

While these apps can also be found on the play store, they aren't as fast or as efficient as native apps built with Java or Kotlin.

IOS, on the other hand, has Swift and Objective C for native languages and the same concepts apply on native vs other kinds of apps.

OTHER HELPFUL RESOURCES

StackOverflow
FreeCodeCamp
Medium

READ OTHER 50 THINGS TO KNOW BOOKS

50 Things to Know to Get Things Done Fast: Easy Tips for Success

50 Things to Know About Going Green: Simple Changes to Start Today

50 Things to Know to Live a Happy Life Series

50 Things to Know to Organize Your Life: A Quick Start Guide to Declutter, Organize, and Live Simply

50 Things to Know About Being a Minimalist: Downsize, Organize, and Live Your Life

50 Things to Know About Speed Cleaning: How to Tidy Your Home in Minutes

50 Things to Know About Choosing the Right Path in Life

50 Things to Know to Get Rid of Clutter in Your Life: Evaluate, Purge, and Enjoy Living

50 Things to Know About Journal Writing: Exploring Your Innermost Thoughts & Feelings

50 Things to Know

Website: 50thingstoknow.com

Facebook: facebook.com/50thingstoknow

Pinterest: pinterest.com/lbrennec

YouTube: youtube.com/user/50ThingsToKnow

Twitter: twitter.com/50ttk

Mailing List: Join the 50 Things to Know Mailing List to Learn About New Releases

50 Things to Know

Please leave your honest review of this book on Amazon and Goodreads. We appreciate your positive and constructive feedback. Thank you.

www.ingramcontent.com/pod-product-compliance
Lightning Source LLC
Chambersburg PA
CBHW020608220526

45463CB00006B/2498